VENUS

MARYSA STORM

BLACK RABBIT BOOKS

Bolt Jr. is published by Black Rabbit Books
P.O. Box 227, Mankato, Minnesota, 56002.
www.blackrabbitbooks.com

Michael Sellner, designer and photo researcher

Names: Storm, Marysa, author.
Title: Venus / by Marysa Storm.
Description: Mankato, Minnesota : Black Rabbit Books, [2023]
Series: Bolt Jr. space explorer | Includes bibliographical references and index. | Audience: Ages 6-8 | Audience: Grades K-1 | Summary: “Suit up! Venus is a planet like no other, and it’s time to explore it through a colorful design that features vibrant photography, engaging infographics, and carefully leveled text”– Provided by publisher.
Identifiers: LCCN 2021022211 (print) | LCCN 2021022212 (ebook) | ISBN 9781623107796 (hardcover) | ISBN 9781644666005 (paperback) | ISBN 9781623107871 (ebook)
Subjects: LCSH: Venus (Planet)–Juvenile literature.
Classification: LCC QB621 .S763 2023 (print) | LCC QB621 (ebook) | DDC 523.42–dc23
LC record available at https://lccn.loc.gov/2021022211
LC ebook record available at https://lccn.loc.gov/2021022212

Image Credits

123RF: leeser, 12–13; Alamy: Buradaki, 10–11, 12–13, 20–21; md sumon, 4–5; UPI, 8–9; European Space Agency: ESA, 16–17; NASA: JET PROPULSION LABORATORY, 10–11; Science Source: Detlev van Ravenswaay, 18–19; Walter Myers, 10–11; Shutterstock: AstroStar, 4–5; Bikomins, 3, 24; cihanyuce, 20–21; Limbitech, 22–23; Mopic, 14–15; Vadim Sadovski, Cover, 1; Vectorfair.com, 14–15; Withan Tor, 6–7

Contents

CHAPTER 1

Meet the Planet

A family **gazes** into the night sky. Far, far away, something bright shines. It's not a star, though. It's the planet Venus.

gaze: to look at someone or something in a long steady way

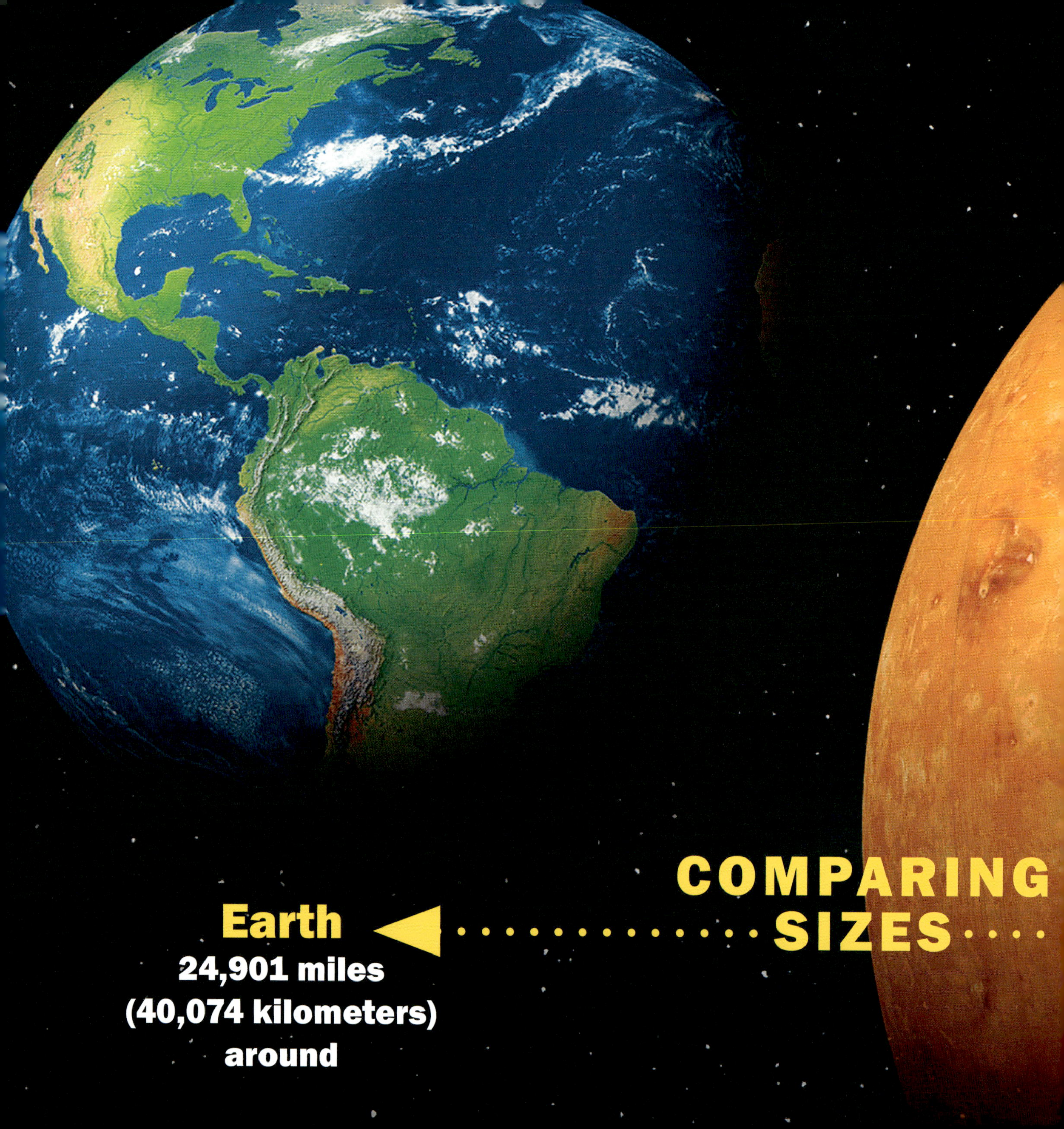

COMPARING SIZES

Earth

24,901 miles
(40,074 kilometers)
around

Bright

It's easy to see Venus. The planet is almost as big as Earth. It's also bright. Sunlight bounces off its clouds.

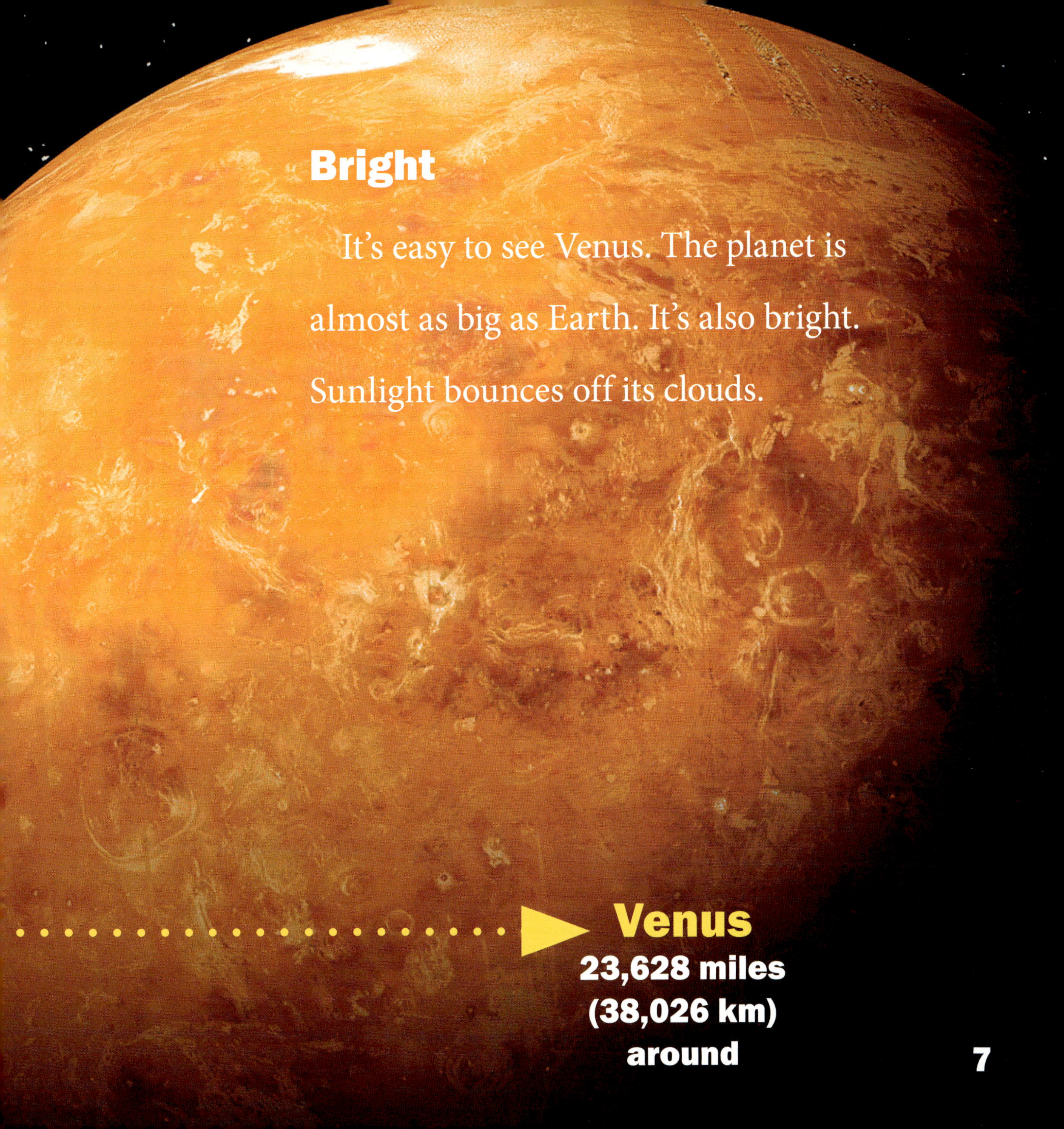

Features

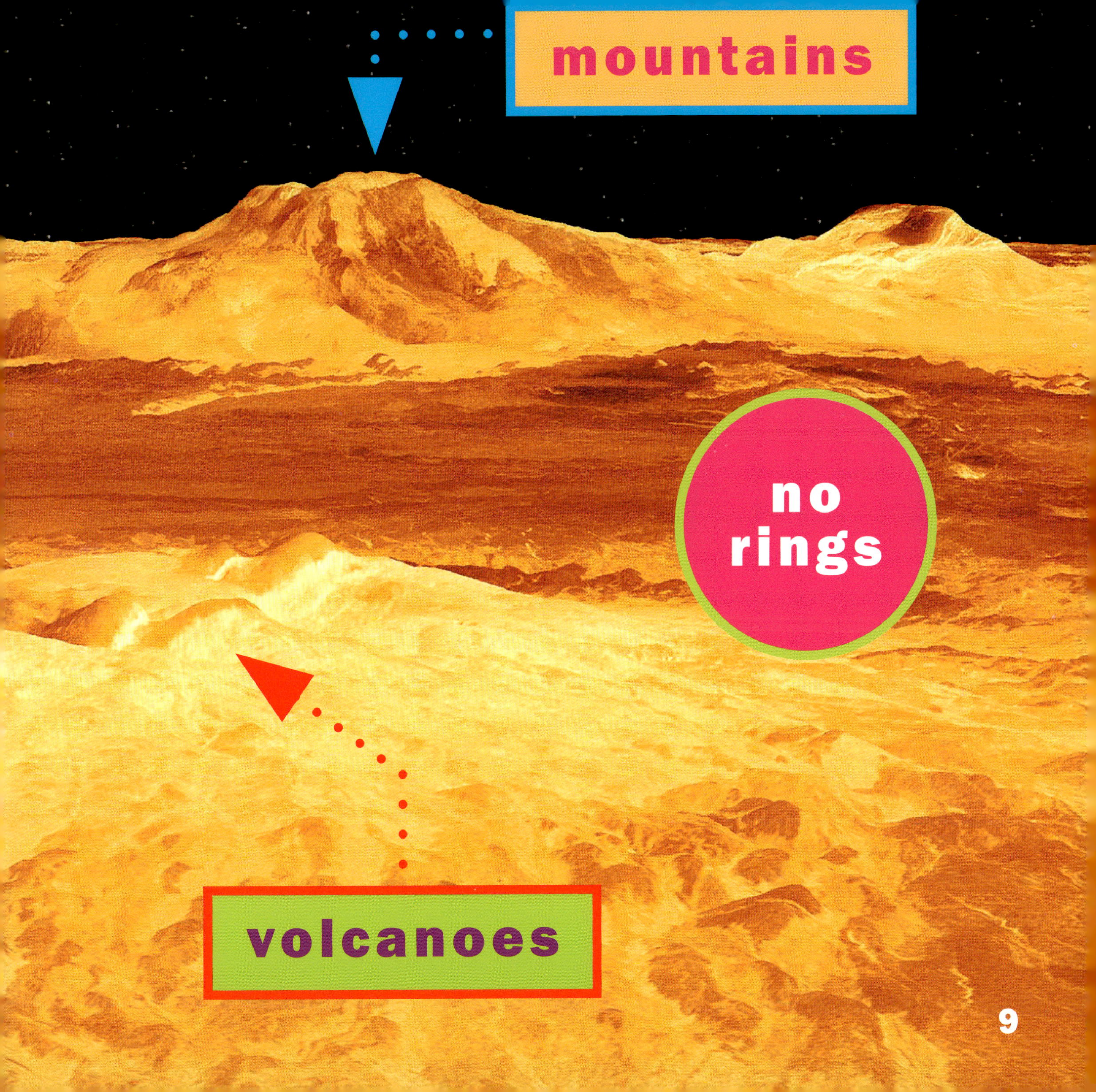
mountains
no rings
volcanoes

CHAPTER 2

Its Place in Space

Venus is dry and dusty. It's covered with **plains** and mountains. It has many volcanoes too.

plain: a large area of flat land

FACT

There are more than 1,600 big volcanoes.

Superhot

Venus isn't the closest planet to the sun. But it is the hottest. The planet has a thick **atmosphere**. It holds in heat.

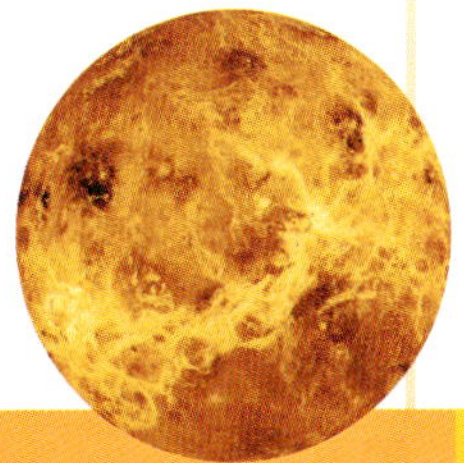

atmosphere: the gases that surround a planet

Venus in the Solar System

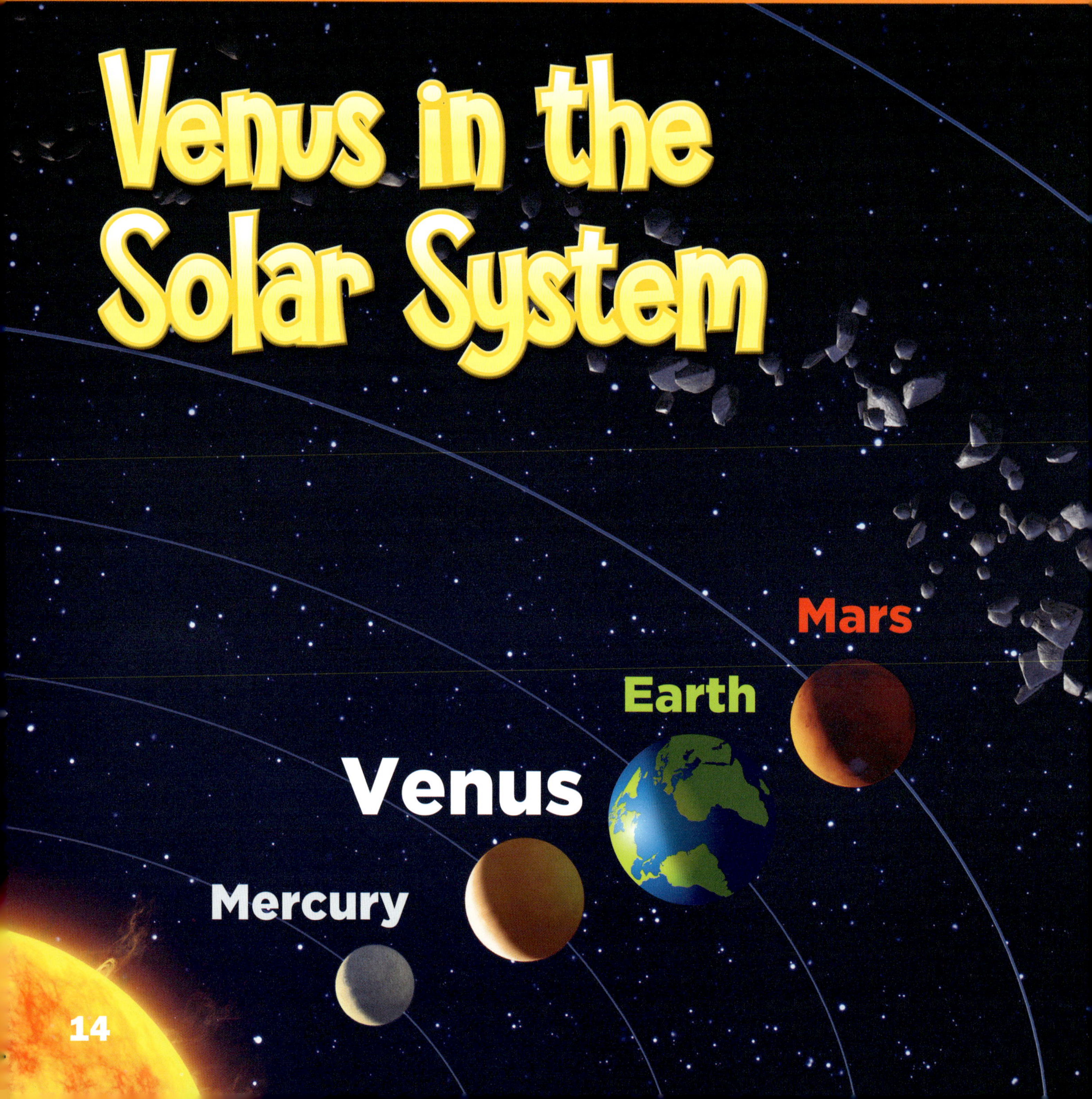

Neptune
Uranus
Saturn
Jupiter
asteroid belt

CHAPTER 3

Explore!

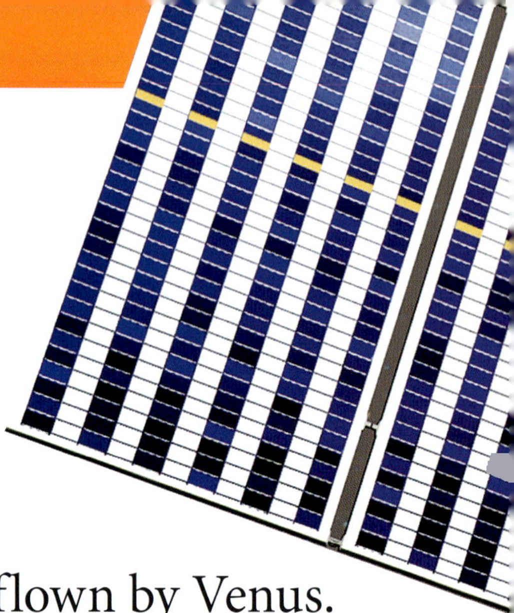

Many spacecraft have flown by Venus. One currently circles it. It studies the planet's weather and volcanoes.

FACT
The first craft flew by Venus in 1962.

esa

The Future

Spacecraft have also landed on Venus. They didn't last long. The heat destroyed them. But scientists keep working. Maybe someday, a craft will survive.

how hot Venus gets

864 degrees Fahrenheit (462 degrees Celsius)

Bonus Facts

Venus doesn't have a moon.

Its winds reach 224 miles (360 km) per hour.

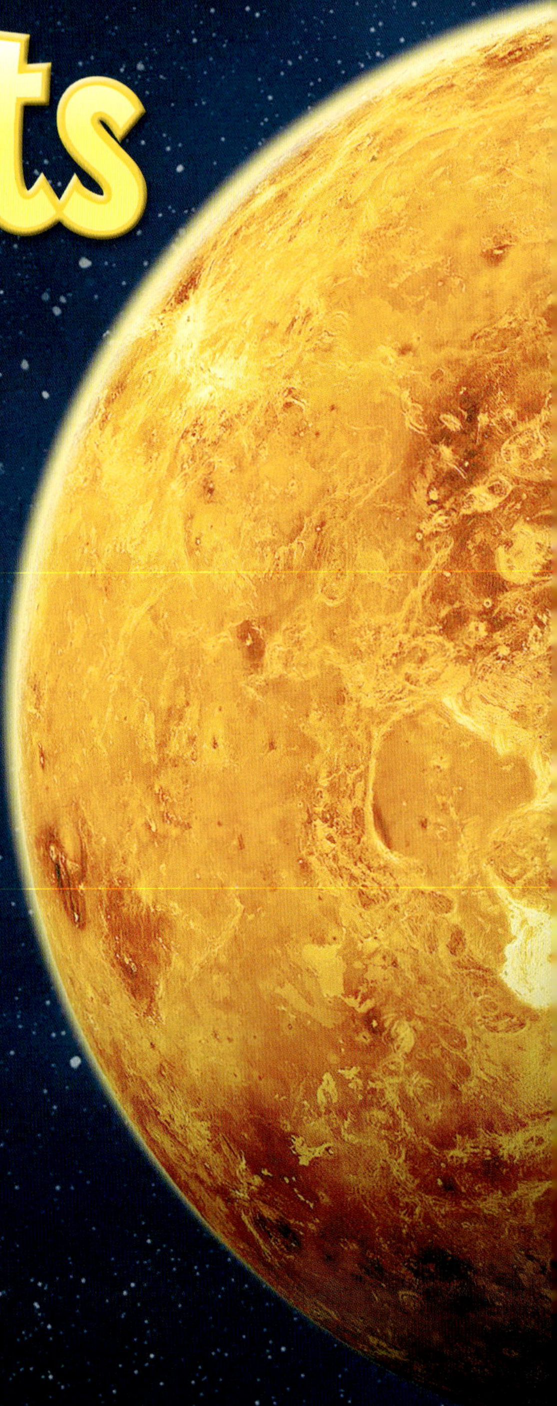

The planet spins slower than any other.

It is hot enough to melt **lead**.

lead: a metallic element that melts at a high temperature

READ MORE/WEBSITES

Boutland, Craig. *The Inner Planets.* Space Detectives. Tuscon, AZ: Brown Bear Books, 2021.

Finan, Catherine C. *Planets and Moons.* Space. Minneapolis: Bearport Publishing, 2022.

Grack, Rachel. *Curious about Venus.* Curious About. Mankato, MN: Amicus/ Amicus Ink, 2022.

All about Venus
spaceplace.nasa.gov/all-about-venus/en/

Mission to Venus
kids.nationalgeographic.com/space/article/mission-to-venus

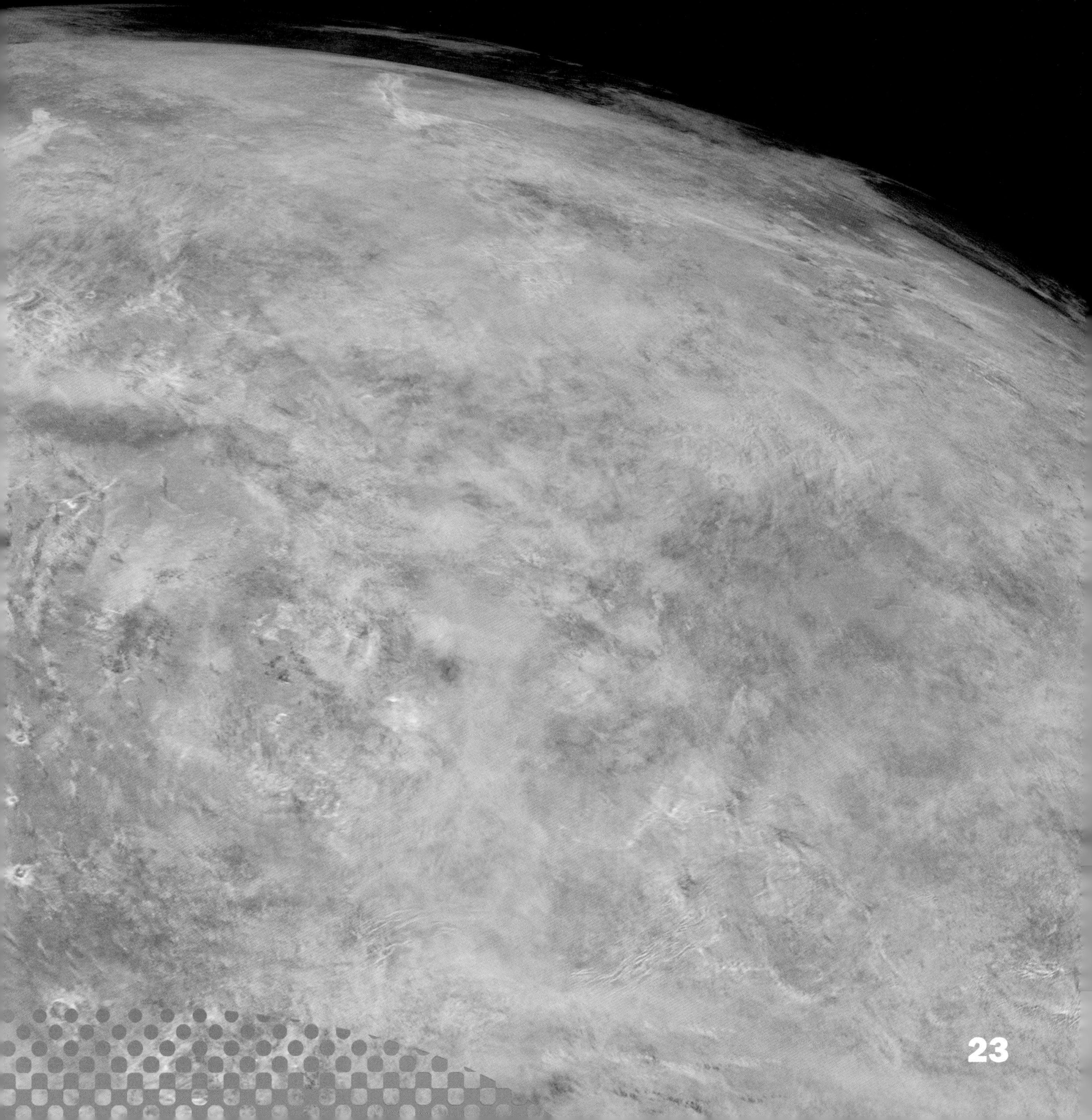

GLOSSARY

atmosphere (AT-muhs-feer)—the gases that surround a planet

gaze (GEYZ)—to look at someone or something in a long steady way

lead (LED)—a metallic element that melts at a high temperature

plain (PLAYN)—a large area of flat land

INDEX